Agrihood Baltimore:

Community Collaboration and Cleaner, Greener Foods

Santana Alvarado

The Facing Project Press

THE FACING PROJECT PRESS

An imprint of The Facing Project

Muncie, Indiana 47305

facingproject.com

First published in the United States of America by The Facing Project Press, an imprint of The Facing Project and division of The Facing Project Gives Inc., 2023.

First paperback edition November 2023

Cover design by Shantanu Suman

Library of Congress Control Number: 2023947237

ISBN: 979-8-9860961-7-9 (paperback)

ISBN: 979-8-9860961-8-6 (eBook)

Printed in the United States of America

10 9 8 7 6 5 4 3 2 1

CONTENTS

INTRODUCTION

"The farmer is a born philosopher, the aristocrat has to learn how."
- Polish Proverb

There are thin places all over the world. Sacred sites where the veil between this world and the eternal world is thin. Spaces where we can walk between both worlds and experience transformation. Plantation Park Heights Urban Farm in Baltimore City is a thin place where time stops, nature prevails, and the seeds of hope and radical hospitality sprout uncontrollably within you, until you're forced to let them see the sun by donning a massive smile.

As project manager and editor, this one-year process was emotional and dynamic. The Center for Religion and Cities (CRC) has supported several of the farm's initiatives and is currently in the weeds of understanding the role listening has in community engaged research. What better way to practice deep listening in public scholarship than through the Facing Project? Initially, I met with farm leaders Tiara, Ayodele, Imani, and Bria to discern whose stories we would highlight. Plantation Park Heights is full of bright stars with lessons we can all learn from, so it was difficult, but we chose folks that represent the farm's diversity.

We collaborated with Dr. Samia Kirchner and her *Design and Human Behavior* students at Morgan State University. Dr. Kirchner, along with writing fellows, enthusiastically interviewed junior farmers, volunteers, farm leaders, and elders. We sought to unpack their life experiences to better understand how they came to join the farm and why they stayed. Editing required further interviewing to adequately capture the essence of each storyteller. I was challenged by the project's substantial scale and its unyielding timeline. While editing, I grew to accept the stories would repeatedly bring me to tears, no matter how many times I read them. As I offer them to you, I am overjoyed by the transformative process I've undergone as a result of meeting these storytellers and spending invaluable time at Plantation Park Heights.

The book's three parts are the farm's core values: *Love, Loyalty, and Respect.* Part one begins with farm president Tiara's mission to accept Baltimore youth as they are while detailing the farm's philosophies and initiatives. We then meet the pepper king himself, Tevin, who recounts how they've fostered the part of him that wants to try new things and give back to his community. Imani shares how growing up with her family's matriarchs and attending Howard University and Johns Hopkins shaped her desire to dismantle our oppressive healthcare system. Ms. Margie's is a compelling story detailing the farm's role in redeeming Park Heights' violent past and restoring its cultural value of community care. Elijah reminds us that if Plantation Park Heights wasn't doing the real work of uplifting the community, he wouldn't be working there. Part one ends with Thomas championing the ideas of shared labor and resources, nature-focused arts education, and radical love.

We start part two with Liam's journey from Trinidad to Baltimore and the hard truths he had to face to become a leader at the farm. Jaliyah's story emphasizes the science of plants and how valuable the weekly free food boxes are to families in need. Bria's powerful story illustrates how the farm's healing effect on her life, after police-inflicted trauma, motivated her to pursue a newfound passion.

Isha shares her pride in being a four-year-old farmer and, at the age of twelve, contemplates a career in agriculture. From Jaylen's story we see the positive impact the farm has on young people's lives, inspiring them to grow in new ways. Samia ends part two exploring the powerful relationships she's cultivated between the farm and the university, paying special attention to the future of design and urban farms.

Shawna kicks off the book's final part attesting to the life-giving influence the farm has had on her neurodiverse family. Khalia then recognizes Baltimore's food apartheid and food sovereignty as a city and regional planner. Harold chronicles how the Center for Religion and Cities was able to support the farm during the pandemic and what he's learned from his friends Chippy and Tiara. In middle school, TJ proclaims he'll never abandon his beloved Baltimore, and will instead stay to uplift the next generation, just like the farm taught him. Finally, the book ends with our delightful founder, Farmer Chippy, as he shares his journey of creating the farm and his dreams of taking this thriving business global. You'll notice both Farmers Tiara and Chippy have twenty-one questions following their stories. We included them because these two are the heart and soul of the farm and we couldn't stand the idea of readers not fully grasping their humor, wisdom, humility, and intoxicating personalities.

This book details how this urban farm, in an alley in Park Heights, has transformed the lives of seventeen community members. The storytellers are diverse in their hometowns, ages, ethnicities, beliefs, education, and personalities but one thing persists: their love for and dedication to Plantation Park Heights. My hope is that this book serves as a tool for more funding and as a model for universities embracing community-engaged research to transform education in alignment with the gifts, needs, and aspirations of their surrounding communities.

I want to give a special shout out to the farm's Arts & Humanities Program Coordinator and Vice President Ayodele LaVeau, their Marketing Director and Project Manager Karma Francis, and my supervisor at the Center for Religion and Cities, Rupa Pillai. These three fierce, hardworking, and steadfast women have not only helped me carry this book to the finish line, they made me laugh while doing so. Congratulations Plantation Park Heights, you earned this book and I hope it only serves you as you continue making all of your dreams come true!

- Santana Alvarado

PART I: LOVE

Farmer Tiara Matthews: The Mentor

As told by Thadius Hodges

I never thought much about heroes, well, until I became one.

Have you ever heard the saying, "Life is like a box of chocolates?" You know the famous line from the movie, *Forrest Gump*. Oh dear, let me finish the quote, "Life is like a box of chocolates, you never know what you're going to get." My name is Farmer Tiara. I was born and raised here in Baltimore, Maryland. I'm the president and farm manager of Plantation Park Heights Urban Farm but most importantly, I'm one of the mentors. Serving youth is my passion. This is a safe space for young people to develop free from fear because there's no judgment being passed here. This space is sacred, this is a space where dreams manifest.

The story of the farm really starts with Farmer Chippy and his friends right next door. Park Heights is a community that has a large Caribbean diaspora so there's a lot of Jamaicans, Trinidadians, Haitians, Grenadians, and Guyanese people who live here with their children. There is a history of Caribbean and African American relationships so we have generations of mixed African American and Caribbean American children. When I started coming here, all of the land that Plantation Park Heights Urban Farm is on today was full of vacant homes. Our contractors, who live right next door, are Chippy's friends and they have been living in their home for thirty years. The issue was that they couldn't get fresh food in the neighborhood and, being from the islands, they wanted to have fresh oregano, thyme, tomatoes, and cucumbers, so they created a box garden right here. Then, Chippy started involving the children.

Our first set of youth were young when they started coming, ranging from kindergarten to the eighth grade. All of the small children would visit, but the teenagers weren't coming up here yet. Then COVID brought a huge wave of youth to the farm. Once they heard about our camp night and that we cook every day, all of the teenagers came, which the younger children weren't too happy about. After a while I started to see more and more youth. It warms my heart that they are choosing to be here as they learn, heal, and grow.

3

To me, children are like plants. They both require love, nurturing, and guidance to grow and blossom into something beautiful. When plants are sick, I nurse them back to health by identifying the problem, adjusting the lighting and watering as needed, removing dead or damaged foliage, fertilizing, and treating for disease. I take pride in teaching kids to be better than their circumstances. What people deem "troubled youth," I don't. I don't think there's such a thing as troubled youth. A lot of them might not be troubled, they just need a place to breathe and a space to clear their minds. The trouble is not with the youth, it's with the system. You are provided a circumstance and you work with it. You either get busy or you fall. If there's no food stamps or welfare, and if you're not working or growing your own food, you don't eat. So you can't blame this on children, they're not making this up. They're just creating their own work and managing their own resources. And what about what's inside? The ancestors. Their spirit. The ancestors that carry over into you. The ancestors you haven't even met. They're still coming through. I think about that when I think about my attitude and how I am. I always wonder how my ancestors were moving because I'm a rebel.

At Plantation Park Heights, we believe that teaching children about agriculture sets them up with a foundation for a healthy life. For example, we teach the youth how to grow fruits, vegetables, herbs, and we even teach them how to raise chickens. Through agriculture, children learn how to navigate life. They learn how to raise plants from seeds and package it up for selling at our local farmers' markets. They even get to make their own foods, products, and seasonings. They learn responsibility. Here at the farm, we love to see the youth thrive as nature brings a sense of calmness to their lives. At the farm they can be relieved of the everyday stressors and distractions that surround them. I've watched how nature has helped them be more creative and heal physically and mentally. Nature is my friend because we both aspire to help children become healthy.

We are building this farm for youth who are obese because they don't have access to healthy food. It's amazing to see how, when children first come here, they usually want fast food like chicken wings and fries. But as we teach them about healthy eating using ingredients we grow here at the farm, their preferences shift toward home-cooked, nutritious meals. We're fighting against diabetes, high blood pressure, and obesity. These are the top three causes of death in Baltimore with homicide and violent killings ranking lower on the list. All of the top ten causes of death are food or lifestyle related, like lung cancer from smoking. Stress too! It's physical and mental health. It's not just what you eat, it's how you live your life. I'm confident that these diseases better be prepared, because the youth will prevail. We have thriving five foot kale and it's because of our youth. When they plant, they do it with love and intention. We must speak life into our youth the same way we speak life into our plants. Our thriving, beautiful, and fruitful plants are just emulations of our youth.

You see we have no fences here. If you go to any other farm or garden in Baltimore City there is some kind of fencing all around it. This is an open space. People ask us all the time, "don't people in the neighborhood steal?" No! We're a family around here, people know that if they want something they can come and get it. They don't have to steal it, it's here for them. Currently, we're leading a "Thousand Fruit Tree Challenge," where we're planting a thousand fruit trees in the Park Heights neighborhood. We're targeting areas by the schools so that when students get out for the day they can pick fresh fruit. The youth take ownership over the farm and our projects because it takes a village.

Even this house on the corner, we already bought it and we just finished the roof. It's going to be a shelter for homeless youth and young people in need. This will be a safe space for all of the youth who are going through things with their parents. We can reach out to the parents and say, "they're here, they're going to stay here for a few days, they're fine." We are providing a model of transitional living *for* the community *by* the community and the city is actually giving money for programs like this. The resources are there, you just have to know about them.

It has taken a lot for us to get to this point. We've lost kids to gun violence, which is so sad and unfortunate, but Maya Angelou said it best "*Still I Rise*." However in this case it's still *we* rise! I'm here to break generational curses, to teach innovation, and to show the youth there is a different way. I'm here to tap into the negativity and show them there is a positive and safe way out.

I want to challenge you. Are you up for it? Starting today, I want you to step up and be the mentor the youth are yearning for. They need positive leaders. They need someone who believes in them and speaks life into them. Let us encourage them to be the best possible versions of themselves they can be. What I'm asking isn't easy, it may be one of the hardest things you've done, but it's needed. The youth are the future. They are our future presidents, future mayors, future teachers, future mothers and fathers. Let us show them the way. Will you take my hand and accept the challenge?

21 Questions with Farmer Tiara

1. **How has the farm contributed to your spiritual life?** It has contributed greatly to my spiritual life because it gives me a place where I feel free of fear and free from fear. It's a safe space to connect with all people from different spiritualities and walks of life. I connect with insects, plants, and animals so it's really allowed me to just be myself and be calm, cool, and collected. Just to have that safe space to breathe in.

2. **What are your core values and where do you see them at the farm?** Loyalty and consistency. Without those two it's

6

really hard to be a farmer because farming is not about making a whole bunch of money or being rich in material things, it's actually about being rich in spirit, getting up everyday, putting in the work, being dedicated to community, families, and children. For consistency, just doing it every day. Whatever you love, you have to do it every day. It's not about having marks of what is good, what is bad, what is the best work, what is the worst work, but just getting up consistently and doing a little bit toward your goals every day. That's what matters the most.

3. **How did you meet Farmer Chippy and what made you trust his vision?** So I met Farmer Chippy hanging out on another block in Upper Park Heights, on a Caribbean block. I grew up around Trinidad-American and Trinidad-Caribbean people living in Baltimore City and I met him at a Trinny block lime (hang out). He has impacted my life tremendously. He is an amazing man, mentor, lover, friend, confidant, trainer, and parent.

4. **If you could give young people in the world one piece of advice, what would it be?** Only one? For me, it's really about staying true to yourself but making sure you don't stay true to the selfish urges that sometimes bubble up inside of us, especially living in a world that is so self-absorbed where everything is about me, me, me. Instead, really indulging in who you are, your family, your ancestors, healing generational curses, and just falling in love with yourself without things. You with your dirty nails, your dirty toes, your nappy hair - love everything about yourself, don't try to conform to the world.

5. **If you could give old people in the world one piece of advice, what would it be?** I don't really have any advice for older people for real, I feel like they've lived through so many testaments in their life that they are where they are. I would just say maybe listen to younger people a little bit more. Maybe be a little more understanding of young people and the

circumstances they're living in today as opposed to comparing it, so you don't become the very people you hate. Thinking of all the atrocities our ancestors and elders went through and still go through. Let's not impose those same atrocities on the youth by trying to control them and making them adhere to a system that never loved us anyway.

6. **What do you think your family, ancestors, and mentors would think of you now?** They would think Aṣẹ, Aṣẹ, Aṣẹ!

7. **How do you encourage creative thinking at the farm?** By letting people have a free space to be able to develop and create on their own whether that's just creating a conversation, a vibe, artwork, or playing some music. We do leadership academies and trainings to cultivate ideas from youth that they may have never thought about but once they start to ponder the ideas, they go into another world they actually enjoy and love.

8. **What would you say your superpower is?** Nurturing, yeah I think my superpower is nurturing.

9. **What is the most important characteristic of a leader?** Being honest at all times.

10. **What mistake do you see leaders make frequently?** Not being honest. Lying!

11. **What's the most important risk you took and why?** Leaving my job and just jumping into urban farming without having any real financial stability but believing in my hard work, myself, the universe, my ancestors and linking up with my partner Farmer Chippy and coming together to lock in and dedicate ourselves to each other and this work.

12. **What are your hopes for the future of urban agriculture?** Oh my gosh, I hope the future of urban agriculture is a place

for Black and brown youth, especially Black boys to find their space to lead the nation. As the old proverb says, "you'll need a lawyer one time, you'll need the police one time, you might need a doctor one time in your life but you will need a farmer three times a day." Just letting them know the importance of food and showing up in the industry that we have led for generations. Slavery did not create the plantation, it did not create the farm. Black and brown people been about that life.

13. **What misconceptions about farming would you like to dispel?** I would like to dispel the misconception to Black and brown people that agriculture and farming ties to slavery which made us get away from knowing what we put in our bodies, it made us disconnect from the earth. That's a trauma we have to heal from to go back to the natural place that feels most comfortable to us and that is farming, being outside, connecting with the soil, the animals, and insects. We have to get back outside and get back to the land. That's where we're most natural at.

14. **What motivates you to keep going even when times are tough?** The young people in Baltimore City motivate me to come here every day. Even when I'm having a bad day, when I come here I be in tears crying laughing because we joking, having a good time, just enjoying life for real. The young people, God, my ancestors, all of that motivates me everyday. My ancestors because they went through way more atrocities than what I gotta deal with today to make stuff jump so if they could make it jump with all of the obstacles and hurdles against them, it's nothing for me to come out here and do this work everyday.

15. **What are your current priorities in this season of your life, and why?** Farmers' markets, leadership, training youth, planting, getting ready for our last crop of the season and getting ready to get up outta here and work on some ag projects internationally.

16. **What makes your life feel purposeful?** Just being. Just having the opportunity to be and to do something makes me feel purposeful. Waking up everyday and having another chance. I don't really need a thing to make me feel that way because when you live in a place like America and you live in a place like Baltimore City, you don't need purpose to just live and enjoy your life and to get up and do it. You see so many people dying, you see so many atrocities going on that you just feel blessed to get up and be in a safe space, a space where you can be free of fear and free from fear.

17. **Describe your perfect day.** Harvesting some herbs, chilling on the farm with my youth, liming, giving Farmer Chippy a big kiss on the cheek, rolling up a big spliff, hopping on a flight, coming back, and doing the work every day.

18. **What's one of your favorite childhood memories?** Oh I have so many, oh my gosh. What is one that really stands out? My first trip out of the country to the Bahamas. I just remember the whole experience, I can even close my eyes and see my first time looking down at the ocean water and seeing fish swim around my legs. Having that international experience is real key for my life and young people being able to travel and meet other people from different walks of life is key.

19. **How do you know when you can trust someone?** That's a good question. I feel like that comes with time. You have to feel safe with people to trust them. As much as people make you feel safe, the more you'll trust them. A lot of times we get council people and Baltimore City leadership hitting me up because I'm known to be able to bring out a large crowd of young people but they're like, "why do they trust you?" It's because I accept them for who they are, I don't judge them, I let them voice their opinions. I still say what I wanna say and I'll still tell them when they're wrong but I give them a space to be themselves without wagging my finger. And I like to have fun too, I'm not uptight!

20. **What's your favorite food?** Shrimp roti.

21. **If you were stranded on a deserted island, what three things would you bring with you?** I would bring a helicopter, a helicopter driver, and a blunt to get outta there.

Tevin Triplett: The Pepper King

As told by Ibrahim Pride

I'm the Plantation Park Heights Urban Farm Pepper King and this is how I got that name. One day, in July of 2020, I was out with my friends riding around Park Heights on my bike. I usually ride about twice a week and this day it was pretty nice. All of a sudden, I came across an alley I'd never gone down before. I decided to ride down the alley and I found chickens. I was surprised and pretty excited that I saw them because I'd never seen chickens in a chicken coop before. There might be some deer or foxes in the neighborhood but not chickens. As I looked around I noticed there was more. It looked like a bunch of plants were growing out of planters on every side of the alley and there was a big round tent in the middle. My friends stopped to find me, we looked around, and discovered it was a farm.

Farmer Chippy was the first person I met at the farm. I remember he was really excited and had a lot of energy. He told us to put our bikes down by the chicken coop while he gave us a tour. Right after, I met Farmer Tiara and she was very energetic too. She said to come back anytime we wanted. I don't know what made me start coming back, maybe the people. I visited a couple of times and now I come often, mostly on Saturdays and sometimes after school. I'm the only one of my friends that still does because most of them moved to other neighborhoods, but they still visit because they don't live far. Right now, I'm in the 7th grade and I turned twelve on August 13th, 2023. I've been coming to the farm since I was eight and I still live in the neighborhood. Even when I go to high school, I'll come because this is a kind place, a place of peace where there's no drama.

My favorite spot at the farm is the hill because it's usually quiet and sometimes I like to be alone. My favorite thing to do besides look at the chickens is to help grow the peppers. First, I learned how to water plants. Then, about two months later, they started teaching me how to plant seeds in trays and one month after that I started planting small plants in raised beds. One day, I planted peppers. Everybody here started calling me the pepper king because I knew most of the names of the peppers. Eventually, they grew and I got to harvest a lot of them so now I'm the king of growing peppers. Along the hoophouse, we grow rows and rows with all types of peppers, including some really spicy ones. The farm even has a hot sauce coming out soon! I like coming to the farm and I'm glad all the farmers taught me how to do what they do.

The farm changed how I eat because I'm more willing to try things. At first, they would offer me food but I wouldn't want to taste anything new. Then I saw I was around people who liked to try new things so I started to try them myself. My favorite dish to eat at the farm is the fish and bake. The farm keeps me out of trouble just like my parents want, not that I would be in trouble, but my parents always tell me to stay out of it and I've come to like working on the farm. Even my parents have come to the farm and when they first did they started having conversations with Farmer Chippy and Farmer Tiara about taking me on field trips. They've taken me to see other farms and we've even gone on out-of-state trips. The last trip I remember was a fruit farm that had all kinds of fruit. It was about 60 acres of land, probably more.

One day, I want to have a key to the chicken coop and the shed so I can get stuff and be able to teach other kids how to plant and help out at the farm. If I could, I would build a shelter for homeless people, bring in fresh produce from other states like we do with the Thursday food box giveaway, order DoorDash to all the houses on the block, and give a thousand dollars to each family on the block. I'm proud and happy with how Plantation Park Heights gives back to the community and helps feed everyone in the neighborhood. It has made me want to find ways to give back too. If I had to convince someone to come to the farm I would say that there's good food, there's nice people, and they teach you how to do stuff that you've never done before.

IMANI BOYKIN: FOLLOWING IN MY GREAT-GRANDMOTHER'S FOOTSTEPS

AS TOLD BY ABRIA MINOR

Before arriving, whenever I shared I was moving to Baltimore, I heard all the stereotypical warnings about high crime rates and abandoned buildings. Being at the farm has shown me a completely different side of Baltimore that people don't normally talk about.

I was born and raised in a single-parent household in the African American communities of Liberty City and Miami Gardens in Miami, Florida. I grew up with a crowd of family members including my great-grandmother, grandmother, mother, aunts, uncles, and a whole bunch of cousins. When my mom went to work, my great-grandmother would babysit me. She would sit me in her garden while she did all her typical work: pulling up weeds, transferring plants she bought from Home Depot, plucking mangoes and oranges off the trees, and watering everything. I learned how to walk by following my great-grandmother around her garden. She had raised five kids by herself, so her garden was a way to feed her family while saving money on groceries. Seeing how gardening was so pivotal for my great-grandmother's health taught me that you don't need to be part of a corporate healthcare system to take care of yourself. Walking and tending to her garden was my great-grandmother's exercise and seeing her flowers and produce grow improved her mental health.

When living with my grandmother, I used to sleep in the same bed as her and watch NBA basketball games, NFL football games, and the Olympics. That's how I got into sports medicine. I was good at science, getting A's in middle and high school. I loved the human body so I thought I just would stick with that interest. Learning about your body and what exercises and foods work for you - that stuff is fun to me. I graduated high school in Miami and decided to further my education at an HBCU (Historically Black Colleges and Universities). I moved from Florida to Washington DC to attend Howard University, where I majored in sports medicine. Howard University offered me an introduction to my Blackness. It instilled pride in me for my ancestral values. Now at Plantation Park Heights Urban Farm, I see that being in nature is part of us, even if people try to reject that. If it wasn't for Howard University, I don't think I would have connected to the ancestral value of being in nature. To me, it was just something my great-grandmother would do but, as African Americans, we've *been* doing this. We've *been* practicing sustainability and making for ourselves for centuries.

Once I graduated from Howard University I moved to Boston for three years, studying at the Brote Institute of MIT and Harvard. After starting as a clinical research coordinator, I was promoted to community engagement specialist. In those roles, I saw how the healthcare system deters people from holistic care. I worked mainly with cancer patients, and I saw that the mental and physical weight of cancer was a systemic issue. In my role as clinical research coordinator, I read through many cancer patients' medical records, and it was clear to me that people who have enough money to pay for everything out of pocket get the best prescriptions and have a better chance of recovery. Then, there are those who only have enough to afford what their insurance is able to cover and whether they live or die is of no concern to anyone at the insurance company. Things like that are unjust, but we're in America so this is what we're forced to work with. I didn't know the policies that made things this way, but I knew I needed to develop a strategy to help. That's what motivated me to go back to school. I attended Johns Hopkins University for graduate school, pursuing a masters in public health with a focus in health systems and policy.

Transitioning from an HBCU to a Predominately White Institution (PWI) was a complete culture shock. During office hours at Howard, I could talk to my professors about anything, but at Hopkins it seemed like all the professors were really busy with their research and didn't have time for me. Howard was competitive but it was friendly competition. At Hopkins people wanted to know my titles, credentials, and accolades before getting to know me. It felt isolating. Searching for what I had been missing, I googled "urban farms near me" and Plantation Park Heights Urban Farm popped up. I emailed someone about volunteering, and they told me to come to the Wednesday farmers' market. I went, and then I just kept going. After several months of working at Plantation Park Heights, I became the secretary of the Board of Directors. The farm has given me opportunities to be part of something positive and the people at the farm became my family. Chilling and talking is pretty much what my family does in the living room or on the porch. And that's just what we do at the farm, sit around and talk to one another. Sometimes we don't even talk at all, we're just among people and we feel safe.

I love the community feel of the farm. Once, a group of us from Plantation Park Heights attended the American Farm Bureau Conference in San Juan, Puerto Rico. This was my first time hanging with the farmers away from the farm. I was usually busy with school and couldn't always participate in regular farm dinners and excursions. So it was nice for me to see that I could relate to my farm family and hold conversations with them without being at the farm. As Black people, we were the minority at the conference. I wasn't surprised, but I felt disappointed because I know that farming is part of our lineage even if we are under-represented.

When I got to Puerto Rico I had a chance to witness the lifestyle of Indigenous Puerto Ricans. They were family oriented. I appreciated the way that some families I spent time with listened to upbeat music and danced despite the fact that they didn't have many resources. That's like being Black in America. In Miami, they try to pit Black Americans and Hispanic Americans against each other, but we're more alike than we are different. In Puerto Rico, we even passed an urban farm on the way to the beach. I felt more connected to Plantation Park Heights Urban Farm through this experience.

For me, the farm is about fellowship, growth, love, and congregating through shared experiences. There's a lot of growth at the farm, both by the plants and by the people. When I think of love I think about Farmer Tiara, she's the nurturing soul of the farm. Love is expressed by the amount of smiles and the fact that people find it difficult to leave once they've arrived. It's not unusual for someone to say they'll be at the farm for thirty minutes and then end up staying for three hours!

Eventually, I'd like to get into research and collect data to show people why we need more funding for Plantation Park Heights. I'd like to show how the types of foods we're growing are reducing blood pressure, and I want to prove that urban farms are improving community members' mental health. Some people think these facts make perfect sense but others ask where the data is to prove what seems like common knowledge. I hope that I will soon be able to conduct physical and mental health surveys to clearly and scientifically demonstrate the tangible, positive impacts on the community. I want to use my skills to benefit the farm and I know I can push myself to do more. Finding ways to restore ourselves ties into my passion for holistic healing in my community. I am reminded of my great-grandmother and it all comes full circle.

Margie Smith: A Safe Haven

As told by Zion Jalloh-Jamboria

There were a lot of ugly events happening around here and it wasn't just violence. We lost a lot of good friends from sicknesses like COVID and everyday life. It impacts these young kids every day. Our community is made up of young, old, and middle aged people from all different walks of life. There are certain people in the neighborhood who act like glue. They keep everyone together and when you see them down the block or hanging out, you know you're safe. My grandson was like that. Everyone knew him. He was killed in a mass shooting on Park Heights Avenue and Shirley Avenue in August of 2022. People still tell me stories of him helping them out or of his smiling face in the neighborhood. My grandson was respectful and always smiling. He was a lot like me in that way. He wasn't no angel but he was my baby. His name was Kenyon Smith, Yonny to his friends, Simon to his grandmother, and Cuzz to his neighbors.

My mother was widowed before her fortieth birthday and raised eight kids by herself, four boys and four girls, without any help from the government. She taught us manners and told us to do honest work. Her friends were part of our community and we lived in a neighborhood almost like a village, similar to Park Heights. All my siblings and I worked and gave back to the community, which is what my mother taught us. I teach my children and grandchildren to do the same. We all do what needs to be done and giving back to the community is what we are doing here at the farm. This is what Yonny was doing when he ended up in the wrong place at the wrong time. I miss him everyday but I am thankful for the changes that are coming because of his passing. I accept his passing with grace because the last thing Yonny said was there is a higher force. I know he is with God. His passing has brought us together as a village and this garden we are creating brings us serenity. This farm, it's a family. This farm is a key element in how we are trying to change around here. We're trying to make Park Heights a safe haven. We won't be having tons of violence coming through here.

The farm being here has really changed our community. Farmers Chippy and Tiara are huge forces of change in Park Heights. The boys in this area know they can rely on the farmers at Plantation Park Heights Urban Farm. The young men come here to work and hang out and it gets them off the street. I know when I don't see one of my grandbabies, they're at the farm. I don't have any worries. I always have my peace of mind knowing the farm is here. That's different from how it has been in Park Heights in the past. Today, on this street, all the way down to Candy Stripe Park, it is safe. When the kids are outside, all of the neighborhood men will come outside to watch and take care of the kids. This all started with the farm.

Farmer Chippy and Farmer Tiara brought their Trinidadian and Baltimorean culture to our neighborhood and it continues to influence how we come together on this street. For example, when I was growing up, if someone died, the family of the deceased wouldn't have to cook for themselves for at least a month. Everyone would join together to buy groceries or bring cooked food. Well, when my grandson died, everyone came to my house looking to be fed instead. There has been a shift in culture away from the community. I've been living in this house for the last forty years, and somehow in those forty years we lost some of our culture and the way we used to do things to take care of our families and neighbors. The farm brought that culture back to our neighborhood. Every time I come to the farm, there's music, hot food, and fresh produce. There are children running around the older boys and there's good company and conversation. I know that every time I go to the farm, I'm going to leave with a new grandchild calling me "Ma" and there's nothing better than that feeling. When children call me Ma at the market, I turn around and look because I know someone could be calling me. It feels good to know the people around you trust you with the well-being of their children because it takes a village to raise a child. This is how I was raised and the farm is transforming our community back to how it should be. We are people taking care of each other and creating a village to grow old in with the younger generation surrounding us.

My mother would invite the neighborhood kids into the house and feed them. She truly was the mother of the neighborhood. I like knowing that I'm following in her footsteps. It feels great to know I'm just like her and that I can make a difference in the lives of the young people here in this neighborhood. I love all of them, I wouldn't trade any one of them. I have a whole lot of children, about 40 adopted grandkids. I invite everyone to come to the farm and the park because these are safe havens we are creating for our children. They know we genuinely care for them and we are making a difference in their lives and in Park Heights.

ELIJAH STATON: WILLING TO WORK, WILLING TO GROW

AS TOLD BY DARREN MARSHALLECK

Being a young Black man living in Baltimore, I have experienced and noticed things that most people in America do not. Because of this, I have always wanted to help change my community and make it a better place. My name is Elijah and I'm a sixteen-year-old growing up here in Park Heights. Growing up in the area I've always noticed there's a lack of food and resources. There aren't any local grocery stores in this neighborhood. Everyone I know either has to drive to different areas to go to a grocery store or find what they can in our neighborhood.

Noticing the problems in Park Heights, I was attracted to anyone or any organization that was trying to help build up the community. When I was in middle school I knew of Farmer Chippy and would see him around my school. The first thing I noticed about him was that he was dressed like a farmer and I was surprised and impressed by that. In 2018, I decided to work at Plantation Park Heights Urban Farm. Founded by Farmer Chippy after he connected with the Caribbean diaspora in Baltimore, Plantation Park Heights is a community growing food for the people of Park Heights. The farm was named one of the country's top ten innovative farms by the American Farm Bureau Federation. As a junior farmer, I am honored to be able to contribute.

When it comes to joining the team at Plantation Park Heights, I'm still surprised by my journey. I've always wanted to help the community but I didn't know it was going to be through the farm. If I wasn't recruited, I don't know if I ever would've joined, but I'm grateful for this opportunity. When I first started coming up here I volunteered a few times and then I determined I would commit. Working at the farm has made me a better person and I've developed the skills I need for my future.

Working at the farm for almost five years has helped me develop a real sense of community and humility. The role of maintaining and growing different types of produce like purple okra, tomatoes, squash, eggplants, and peppers has made me more responsible and cautious. Maintaining the farm as a whole takes a lot of work and upkeep. Since we're one of the top farms in Baltimore, we always have to be on point. I don't believe we're considered a top farm by luck. As a farmer, I can say that the food we grow tastes better and is more nutritious than food in grocery stores. We have a variety of fruits and vegetables that everyone in Park Heights can enjoy instead of having to find a store by driving outside of our neighborhood. Recently, I was interviewed and said, "They don't have to go to the store if they can just come here." Our food tastes better and we're the most convenient option for our community.

With that being said, there is more to Plantation Park Heights than just our food supply. We have connected several families with resources like eviction and homelessness prevention services and energy-saving subsidies. What I've learned from working here is that the farm isn't just about growing food, it's about focusing on the community. If that wasn't the goal of the organization, I wouldn't be working here.

When you think about people working on a farm you wouldn't think of teenagers or children, at least I wouldn't. For this farm however, that's the case. There are adults, teenagers, and children volunteering here that all help maintain the farm. I know that when people hear about different age groups working together they might think there are problems, but that's not what happens. We're all on one accord which is really important. Different types of people working together makes everything easier. Sharing the workload throughout the farm allows our duties to be done faster. The elders who used to work all day don't have to anymore since there are teenagers, like myself, who are willing to work. Some of my favorite memories at the farm are when I first began working here. Being new made me love and appreciate the process of helping build up the farm. To see what the farm has become makes me proud.

When it comes to the actual produce we grow, people always ask me and my co-workers whether it's good or not. At the farm, we make sure that all of the food is nutritious and we are educated on the importance of it being healthy. We understand that people need food that fulfills them and provides them with nutrients. Our reputation is key and we always want to keep it up. Being a Park Heights native also brings insight into what the community needs, which is why the farm is very successful. Even though the farm is based in the Park Heights neighborhood, I want to expand throughout Baltimore. There are a lot of vacant places where new farms can be developed and established. The farm is currently expanding by connecting with different programs and organizations and I know we can expand more. I am excited to help the farm grow.

I think working at Plantation Park Heights Urban Farm has been one of the best experiences I have ever had. When I am working at the farm I'm making a difference in my community. Without the farm, people in the community would not have access to healthy food and I think we would be more divided. This farm shows us that when people come together with a common goal, they can accomplish it. I would advise anyone and everyone to come visit Plantation Park Heights because if this farm can change my life, it can change yours too.

Thomas Mooring: The Currency of Authentic Connections

As told by Samia Kirchner

I still remember my first conversation with the young farmers of Plantation Park Heights Urban Farm. It was 2019 at a farmers' market and I was called to their table to try some peppers. The boys were literally yelling at me, convincing me to give their produce some attention. That was one of the best decisions I could have made. I didn't live in Baltimore at the time but I worked as an art teacher in the city. Farmers Tiara and Chippy lit up when they found out that I actually worked at a school that some of their junior farmers attended. As the sun began to set, they invited me to join them at the farm's upcoming event. That divinely orchestrated day was the beginning of a familial connection I never could have predicted.

Growing up, centering myself in communal spaces was essential for the preservation of my joy. At family gatherings I was always running around playing some made up game with neighborhood friends. Who I am today is greatly shaped by those interactions and the ways I carry myself in community. I hold myself accountable and am able to hold the people around me accountable with grace, empathy, and a gentle touch. What I value most are times when I am surrounded by family, especially on holidays like Thanksgiving and Christmas.

It's not uncommon for me to travel to the south because my dad's side of the family lives in North Carolina and my mom's side lives in Georgia. Things are a lot slower there and people have more time to gather. In the rural south there's more time for family, more time to share resources, information, stories, joy, and laughter. I miss that. But that regenerative time to connect with others is what the farm is giving me now.

In urban areas like Baltimore, time to gather is stolen from people who have few resources and who have barriers that don't allow them to rest. People need time and capacity to do things that create authentic connections. We have to get creative with how we find our small salvations. Folks who gravitate to the farm are people who have similar values. Values like shared labor and resources, uplifting the

Black youth, and learning from others. Radical love is a value that I hold close to my heart as I navigate the world. Radical love is an action. It is respecting and making room for yourself and others as equals. It's not about tolerance, it's about true acceptance. Radical love reverberates through the connections that are birthed and sustained on this chunk of land in Park Heights. It's a joy to shape and be shaped by these experiences.

I was born in Maryland and it has remained a home throughout my life. Many avenues, addresses, and parks have held me through my evolutions, as well as my implosions. Maryland is a capsule for the majority of my truths. I went to college at The Maryland Institute College of Art (MICA) in Baltimore City. My primary practices are printmaking, painting, and art using graphite and charcoal. I was teaching these disciplines in K-12 classrooms. The city offered me new possibilities for how I showed up in the world, especially compared to when I previously lived on a military base. In Baltimore, I could explore versions of myself that needed air and repairing. In those first years, I explored my sexuality and gender expression, acquired a love for teaching the arts, and unlearned degenerative ways of engaging in community.

Now, I want to reclaim my art and use it for my own expression. I don't want to teach it the institutional way. Teaching art was too mechanical because you're just hammering out art projects as assignments. However, I would like to teach art here at the farm. At Plantation Park Heights, it would be about expressing yourself and owning your art without the judgment of a grade. I would start the lesson with a work of art and ask the young farmers what feelings come up. I would teach them that art is about getting into their bodies. We wouldn't use technology, unless it's part of the project. It's necessary for people to connect with themselves first, through moments of silence and deep observation in nature. Trying to somehow tune out the modern world because the absence of the modern world can allow us to connect with our spirit.

The farm is like a portal. When you're at Plantation Park Heights you're in another dimension. Every now and then, a car passes and brings you back to our modern urban reality, but soon it falls away. When you focus you can hear the sounds of plants brushing against each other, you can smell the fresh crisp air, and you notice just how green the leaves are. Here is where I want to start teaching art. I want young farmers to put their feet directly in the dirt to get in touch with their physical selves and have a chance to use all of their senses without gadgets. My favorite memories of the farm are the campfire nights. We all connect and embody our values of being involved in the lives of local youth and celebrating our labor through communal food preparation and eating. On camp nights all of the adults cater to the junior farmers the entire night. We enjoy serving them because they deserve it.

We also deserve to celebrate with each other and share experiences. Growing food and cooking is a ritual that grounds us in radical love. Eating with a group of people is a ritual. It is spell work. Trusting the energy of the person cooking and therefore trusting the energy of the food is an important part of the process. All that is cooked here is made with love and you can feel that love in every bite. When I think of love I think of Farmer Chippy's famous soup. Breaking bread together is an embodied value that speaks to the richness of both the food that nourishes our bodies and the neighbors that nourish our hearts.

It is absolutely necessary to replicate this urban farm model across Baltimore City. We're not just growing vegetables, herbs, and flowers, we're also keeping the community *to-gather*. We need to prioritize instilling the embodied value of radical love through food and a mindset of abundance. Plantation Park Heights Urban Farm is setting folks up with a different foundation than they were taught. We all know the morals, or lack thereof, of capitalism and we feel it reverberate through everything we learn in society. But at the farm it's not an "I" thing, it's a "We" thing. That is the currency of authentic connections.

I have ties to Detroit, Georgia, the Carolinas, and Pennsylvania. I want to visit different states but I don't think there is one place that could keep me forever that way Maryland has. I want to be somewhere I am surrounded by nature and don't have to search for it. Here in Maryland I have created a chosen family and I want to continue to grow this circle. Here at Plantation Park Heights Urban Farm we are strengthening connections that will multiply exponentially. As long as we are practicing radical love and eating Farmer Chippy's delicious soups, we will generate the currency of authentic connections.

PART II: LOYALTY

LIAM CAMPBELL-TEAGUE: WHAT'S YOUR WORTH?

AS TOLD BY AZADE DIYKAN-HUBBELL

Farmer Chippy sat me down and asked, "Why would I take you back? What's your worth to the farm? Have you increased your worth since you left? What have you done differently?" I will never forget that conversation. It hurt but it was a good kind of hurt because he was right to question me before taking me back. The truth was that I now had to prove myself to Farmer Chippy. Sometimes the truth hurts.

I was born in 1999 in New York. In 2008, at the age of nine, I migrated with my mother to Trinidad, where I lived for ten years. My mother was in search of a fresh start and wanted to show her home country to her children. In Trinidad, I had a good childhood because like everyone else there, I was always outside. There were fruits and vegetables everywhere. Switching from New York to Trinidad was a big change, but it gave me good life experience. The challenge was that I was an American kid, and there were no more McDonalds or Chuck E. Cheeses in my new home. I was far from my father and extended family in New York and had to change my whole life. But living in Trinidad was the best way to embrace Caribbean culture. With Carnival, beaches, and all the warm personalities, you always feel like you're on vacation.

I'm humble because after starting my life in a big city, I learned how to live a good, simple life in a third world country. There were a lot of times we only had bread and cheese to eat but I'm not complaining, it's delicious! My family taught me to farm and live off the land in Trinidad, which was completely different from New York. I remember going outside to grab fruits for juice or herbs for seasonings. In 2019, after eleven years in Trinidad, I moved to Baltimore by myself, a few months before my twentieth birthday. My mom was pregnant and wanted to come to the U.S. so my sister could be born an American citizen. The tickets had already been bought before my mom realized she was too far along in the pregnancy to fly. She asked me if I still wanted to go and I said yes. I decided that I would come to the U.S. because I was already a citizen, wanted a new start, and knew there would be more opportunities here.

Moving was hard because I was alone, but it wasn't my first time in Maryland. In 2015, I had lived in Hagerstown for a year when my mom fell and dislocated a couple disks in her spine. She had opted for surgery at Johns Hopkins Hospital and got six titanium screws and two pins put in her back. My sister and I lived with her in Hagerstown that year while she recovered and then the three of us went back to Trinidad. My mom chose Baltimore because she loves Johns Hopkins and that's where she wanted to have her baby, who ended up coming a month early.

When I first moved, I stayed with my two cousins and the week I flew out, Farmer Chippy came by to pick me up. To help situate me, my mom had called a friend, and her friend had called Farmer Chippy, the founder of Plantation Park Heights Urban Farm. He took me to Whole Foods for breakfast, helped me get an ID Card, and connected me to the Trini community here in Baltimore. Farmers Chippy and Tiara even hooked me up with my own apartment. It was freedom. I gravitated to the farm because of the people, food, and all of it being outside. It became my new spot.

For the first couple of years, I would catch a ride with my cousin to work seasonal gigs at Kohl's. I was already at the farm, but wasn't mature yet. I was still chasing money and wasn't driven to pursue my next step in life. I just knew I needed to sustain myself and earn some money. I didn't have any goals. For about a year, even though I had been given responsibilities at the farm, I stopped coming around. The farm was the first thing I had found when I got to Baltimore. I was new to navigating a chosen family and, after my first year with the farm, I thought I wanted to try something different. So, that year I left and worked for Amazon and a taxi service. I would hang out and work with the cousins I lived with when I first moved to Baltimore. It was with them that I got into working with my hands through contracting work with concrete and sheetrock.

But what the farm had going on still resonated with me on a real level. I liked being with the kids, growing food, and liming, which is what we Trinidadians call hanging out. We make liming a sport. The farm was home, and I started missing it a lot. There wasn't another community for me out here in Baltimore. I told myself, "you have to get over it, you have to return to the farm and see your people. They care about you. Stop hiding."

I returned to the farm and asked if I could come back and be part of it again. Chippy sat me down and asked me critical questions about my worth and the value I could add since leaving. When I told him I had been working in contracting, he encouraged me to go to the North American Trade School, where I learned about electricity, construction, welding, and mechatronics. I did a little bit of everything. I went to trade school because I knew I wanted to work with my hands and Chippy challenged me to do it. After that conversation, I went on a warpath to get accredited and gain skills. A lot of the men that pass through the farm have a trade and I wanted one too. I did the whole thing by myself. Chippy would say "finish school first." I wasn't banned from the farm or anything, but it was about getting in his good graces for real. He needed to know I was serious.

Now, a few years later, Chippy is still my mentor, helping me find what is most important for me in my life. When I came to the farm, my life changed for the better. It was hard but I created a home and a life for myself with the help of Farmers Chippy and Tiara. I now try to be a role model for the children who are here. I didn't always want to be a role model, but it came naturally because I'm always at the farm caring for them. I automatically took on that big brother role. I buy them sodas and I include them, that's really all I do. The problem in the city is that these kids aren't nurtured. When I look at my childhood, I think about how we were always outside. I was raised with Muslim values, where you always help people. Now the kids will come see me first and know they can ask me for help or an Uber ride home.

One of the main reasons I'm here is because this farm is home for me. I have everything I need here, including friends and opportunities to grow financially. My life is fully built around the farm, especially feeding the chickens. My grandfather had a chicken farm in Trinidad so everything has come full circle. I ended up in an urban farm, doing the same stuff I was doing in Trinidad. I sell food at the farmers' markets and Farmer Chippy trained me as a farm chef to cook all the food we serve here. I fry the fish and bake, but my favorite thing to cook is a Caribbean dish called bake and shark. The sauces are really important. The culantro and tamarind sauce together *chef's kiss.* I love cooking for my community. I miss my country, but doing something with Caribbean culture has helped me manage that emotion. Trinidad's environment is different, being closer to the beach, with mountains and lots of green. However, in both places people are always outside, which means communities can come together to cook, help each other, and chill. That's how I adapted to Plantation Park Heights so easily.

Now, I have goals. I no longer believe that only focusing on money will give me a good life. I'm going to apply to the Landscape Architecture Department at Morgan State University because I love being outside. I know the farm, I've learned about infrastructure, and I'm passionate about the land, so landscape architecture will be good for me. I know it won't be easy but I moved to the United States without anything. I had no future plans and no friends, but now I do. I won't stop. This farm has opened my eyes because it offers great opportunities for young people's futures. So, I will do more things for my future and for the future of this farm.

Jaliyah Everett: Garden Paradise: An Extension of Our Backyard

As told by Jenny Umana-Lemus

When I first set foot on the farm the smell of basil, mint, and rosemary filled the air around me. I saw the familiar faces of kids I go to school with and the elders of the neighborhood. Within a few days, they were no longer strangers. Now looking back, I think about what my life would be if my uncle had not introduced me to my garden family. The roots of the farm are Farmers Chippy, Tiara, and Liam (who we call LumLum). They have shown me how to love and care for nature. For the past three years, I have started the day by grabbing the hose to water the plants. I also plant seeds and I talk to them. My dialogue with the plants consists of manifesting their growth. The growth of plants is important, since they will give Park Heights fresh ingredients in future seasons.

Park Heights is a tight-knit community and my friends come to the farm often. When the day of harvest comes, I'm anxious to get to the farm to meet the other junior farmers. I gently remove the herbs or vegetables from their stems. Afterward, I clean the produce in a big bucket to prepare them for a delicious meal which is overwhelmed by joy and laughter. Farmer Chippy prepares a healthy meal using ingredients from the farm. I enjoy learning how to make veggie burgers and I love spending time with my garden family. These are the memories that I remember most when I tell people about the farm.

When younger children arrive at Plantation Park Heights, I try my best to explain to them what horticulture means. There are kids that wander off and get distracted the moment they get here but there are kids who want to know more about the science of plants. For example, how long will it take for a plant to germinate? How much water should we use? When I first visited the farm, I didn't know that overwatering can damage a plant. We must water in moderation, keeping the weather in mind. As a junior farmer, I'm on the lookout for potential predators that can cost us our harvest. There are bacteria that get carried with wind and water that deteriorate crops. Insects can disrupt the nutrient pathways and our plants will stop growing. When I witness these cases at the farm, I immediately tell the farmers. I have to warn them to avoid the insects spreading to other crops. I tell the younger farmers what to look out for like small holes on leaves which means insects have been biting them. The way we resolve this situation is with a yellow sticky card, placed at each crop and used to identify the pests that are invading the plants.

Plants have delicate traits and I have learned that they need their own food. For survival, they need fertilizer composed of nitrogen, phosphorus, and potassium. These are vital for the plants to grow strong and tall with bigger and brighter fruits. There are also plants that have to be in a contained environment. The hoop house is used to keep in humidity during the colder seasons. The black tarp inside the hoop house is used to sustain the soil's moisture and kill weeds. This allows plants to continue growing in Baltimore's cold winters.

The mission of the farm is to give back to the community. Our gardeners prepare food boxes for residents to pick up. I'm proud to say that the time I spend caring for the plants pays off. I love seeing relief on the faces of mothers who did not know how they were going to feed their children. I wish that more kids could experience an urban farm in their community. Especially the youth that come from families struggling to buy groceries. When I see the mothers' gratitude and relief, I think of the women figures in my life. Perhaps there were times I did not know my family was struggling. My grandmother and aunt have worked hard to give me a good upbringing and pushed me to pursue my interests.

Farmer Chippy encourages everyone to express themselves and enjoy their time on the farm. He began the farm by basing it off of his upbringing in Trinidad. Farmer Chippy spent most of his life gardening, allowing him to become familiar with nature and his culture. The herbs that he planted growing up were used for cooking cultural dishes and treating illnesses. Now that he lives in Baltimore, he has shown the younger generations these African and Caribbean traditions and because Park Heights is on fertile ground, many plants are able to grow here.

I have learned how African herbal healing has different health benefits like boosting immunity and treating infections. It's fascinating to know that we can nurture and treat our bodies with the plants around us. My mother began making calming teas at home from the basil and mint we harvest at Plantation Park Heights. The smell of basil and mint intensifies in our home when the water starts boiling. She says they reduce anxiety and it helps me concentrate and sleep better. The farm has helped me stay active and changed my eating habits. Before I became a junior farmer I was eating a lot more processed foods. Now I'm able to speak up in my household about making healthy choices. I'll choose a recipe that I learned from the elders at the farm and my family members will help me prepare the meal.

I wouldn't change anything about the farm. Our gardeners use the tools we have and make the best of it. During the sunny days, we find shade near the trees or use the shadows from buildings to take breaks. There's always seating available to sit down with friends and family. It feels like an extension of our backyard. We manage our resources and teach the community how to maintain this treasure. The goal is to grow more, so that every family can have food security and gain access to our garden paradise.

Bria Morton-Lane: The Great Escape: A Journey of Healing

As told by Lilleana Watson

One day, when I was eight years old, about twenty cop cars rushed toward my home in Randallstown, Maryland, a predominantly black neighborhood located west of Baltimore City. They raided the inside, searching every nook and cranny, including our bedrooms and even the refrigerator, all while my sister and I were home alone. My life changed drastically that day. I had to quickly adjust to a different lifestyle as a result of the raid. Before, I was like most of my neighbors, I lived in a two-parent household with a dad who taught me how to ride a bike and hit a softball in our big backyard. After the raid, I lived with my single mother, frequently moving from house to house trying to find somewhere she could afford on her own.

All my life I have had to be very adaptable to the new living conditions I've been placed in. Even with everything that happened, my mother still expected nothing short of greatness from me in everything I did. This is due to my family's history and William Lee Morton, my great-great-grandfather. One day he was moving furniture in Baltimore City when someone asked what he did on Sundays because his services were needed. He responded by stating, "I spend time with my family on Sundays." After careful consideration, his moving company was born. The *B. Morton Express* moving company was the first Black-owned moving company in Maryland. Eventually, my great-grandfather, Bernard Morton, inherited the moving business which was then passed down to my grandfather, Bernard Morton Junior. I have all this power and knowledge because I learned my family's history. Ultimately, it taught me perseverance and determination. I know how to keep moving despite my circumstances. Since the raid, my family and I have been doing our best to heal from the trauma we went through. In the process of healing myself and joining the farm, I've developed a newfound passion for wanting to help heal others.

I started at Plantation Park Heights Urban Farm during my freshman year at Howard University. At first, I volunteered to spend time with my cousin, Farmer Tiara, during summer break and when I had time off from school. Being at the farm has always felt like an escape from reality, it's a place where I have the world at my fingertips. There are so many different cultures, religions, and lifestyles that I interact with and I always enjoy learning from others. Building relationships with a diverse set of people has helped me understand the true meaning of humanity. What I mean by humanity is that, regardless of where someone grew up or what language they speak, we are more similar than we are different. It was at the farm that I was able to notice the similarities between people of different cultural backgrounds. Initially, I didn't understand the mission or story of the farm until the COVID pandemic changed my relationship with it. Plantation Park Heights went from being an escape to being my life.

When the pandemic hit and we went into lockdown I had to come home from college. I noticed Plantation Park Heights was one of the essential businesses that remained open and continued to serve food to the community. I had more time to myself and ended up spending more time at the farm, helping out with tasks like planting and running our farmers' markets. Since then, I have worked my way up to becoming treasurer of the Board of Directors. I've learned the health benefits of herbs and produce grown at the farm for our different body systems and organs. I've had the opportunity to lead the farmers' markets which involves listening to customers and providing them with the best produce and service. Customers would return each week with stories about how their stomachs no longer hurt because of lemon balm I grew and suggested to them or how their cough was alleviated because I recommended mullein. Seeing their smiles and hearing their testimonies instantly made me forget about the intense labor we put into farming, which I was still growing accustomed to. It was worth the long days planting and harvesting in ninety degree weather, watering plants in a sweltering greenhouse, working from sunrise to sunset, and even being itchy from harvesting okra in short sleeves. My newfound passion for learning how to treat the body and using that knowledge to help treat others has led me to pursue a career as a physician. Currently, I am in the process of applying for medical school. I plan to bring the holistic and spiritual perspectives that I've learned at Plantation Park Heights to the medical field, making it more inclusive for people like me. The farm has not only helped me discover myself and my passions, it has also been a pivotal addition to the Park Heights community.

Plantation Park Heights' mission is to provide healthy food while creating healthier and more sustainable communities. Being here has helped me see the impact that an organization consisting of like-minded and passionate people can have on its surrounding community. A lot of the people who come to work at the farm are similar to me. We come from high-risk situations and we ultimately use the farm as a way to heal from our personal experiences. Growing food is powerful. It helps heal our bodies. Physical health is vital for healing mentally because the mind and body are linked. The farm is helping heal individuals and whole communities. Plantation Park Heights has been a safe space for me for as long as I can remember. I hope the farm can be a safe space for others on a larger scale, a space for people to grow and cultivate their passions as individuals whether through farming, the arts and sciences, teaching, or culinary arts. Everything I have been through as a child, my family's history, and my experiences at the farm have made me who I am today. I will use the values and skills I have learned like hard work, perseverance, and holistic practices to not only be a great physician but a great leader that is always willing to help people in my community heal.

ISHA JOSEPH: ISHA'S URBAN INHERITANCE

AS TOLD BY JAMES TOTTON

I think I know what I want to be when I grow up: a farmer for the Park Heights community. At the age of twelve, I am a volunteer, a junior farmer, and a teacher at Plantation Park Heights Urban Farm. When I'm not planting at the farm, I'm educating young and not-so-young people about agriculture.

Originally from Baltimore and as a girl of Trinidadian descent, I started farming in the very early stages of the farm's inception. Even though I call her Aunt T, Farmer Tiara is my godmother. When they started the farm in 2014, Aunt T always brought me along and I enjoyed it so much. She taught me what it takes to maintain a farm, like planting and harvesting. Volunteering at the farm is an everyday routine for me and it has been since I was four-years-old. I've learned and taught others how to plant seeds and water them, I've given tours, and I answer any questions people have when Farmer Tiara isn't available. Being able to teach people new things is amazing and as an OG here I am proud that I can help people. The farm is full of good experiences and I like that I can be myself.

Outside of farming, I like makeup, fashion, and posting TikTok videos of me and my friends dancing. In the past eight years, I've brought a few friends here and even when we're farming, you'll find us joking, chasing each other around, and making time for peace and quiet. The farm is a place where I can have some alone time in the shade when I need it. While planting, you'll hear relaxing Soca and Reggae music playing from the speaker. If I had to convince my friends to visit I would say, you don't need to come if you're not comfortable but I want you to because people dream about being in places like this. You can build knowledge and have a great career here. Farmer Tiara and Farmer Chippy can help you gain skills, improve how you eat, sleep, and even help get the bumps off your face.

The Wednesday Farmers Market at Druid Hill Park by the Rawlings Conservatory is my favorite. We get to the farm at 10 am to harvest before we leave at 2:30 to set up. The summer is always a busy season for us and I assist the customers when they're purchasing our goods, including homegrown plants and fresh eggs. I love going up to people and talking to them.

   I will definitely continue to come when I start high school because this is my life! Once I even tried giving it a break for a few months but that didn't work out. I ended up coming back after a week. I'm in middle school now but I hope that working on the farm will help me get a scholarship one day. I want to go to a competitive college and I won't be able to if I don't have a good record and experience. I have a whole plan and mission that I want to complete here. The farm is an escape from the harsh realities of the world like homicides, suicides, and poverty. We try to stop that by bringing people here. At a time when the city is ranked one of the most dangerous cities in the United States, it is so refreshing to have a space for the public to volunteer and plant locally grown produce for the community.

   The peaceful atmosphere is part of what keeps me coming back every day. It's gotten to the point that I see myself being in the field of agriculture even though I have plenty of time to think about my career goals. My friends and family recognize the gift I have to educate people about agriculture. For now, I'm enjoying the present moment and experiences I get to share with my community. My dream would be to bring other communities to the farm to let them see how we are persevering and prospering. We spread positivity because you never know what negativity is inside people. You won't get anywhere if you don't believe in yourself. I believe in Farmer Tiara and everyone here at the farm. Everyone can get where they need to be. Farmer Chippy is showing us we can strive for higher than where we are. We can't get there with negativity and not believing in ourselves. When you say "I got this," other communities will come and want to learn.

Jaylen Jones: Farmer, Chef, and Much Much More

As told by Harold Antoine

After somehow stumbling upon the farm, my older brother and I were some of the first kids in the neighborhood to help around at Plantation Park Heights and become farmers for the community. My name is Jaylen Jones and I'm a sixteen-year-old who grew up here in the Park Heights neighborhood of Baltimore. When I was younger, my older brother used to walk past the farm all the time. Eventually he got me to work on the farm because it was something positive we could do. What brought more youth was sharing the idea that they could have fun. For some reason, kids love getting messy and playing in the dirt. The founding farmer's name is Richard Francis, known as Farmer Chippy. He founded the farm after connecting with the Caribbean diaspora in Baltimore to provide food to his Park Heights neighbors.

Working on the farm with Farmer Chippy, I learned quite a bit of information. The farm helps me by teaching me how to grow produce that I can cook with. When learning how to plant and harvest produce, I learned that it minimizes carbon emissions and pesticide use. When you pick vegetables straight from the garden, there's no need for plastic packaging. This farm helps us develop a relationship with nature. Farming offers learning opportunities, improves our nutrition, and the food just tastes better.

I have two future goals. One is to be a professional NBA player and the other is to be a chef. After high school, I want to attend either Bowie State, Coppin State, or Morgan State to play basketball. My favorite basketball player is Zion Williamson, a power forward for the New Orleans Pelicans. Playing basketball has a lot of health benefits and the farmers encourage me to play. Sometimes they even take me to the basketball courts to practice.

Being on this farm has taught me to be more caring and to think of others. I like to help younger kids by showing them the ropes of how to farm. Outside the farm, I help my mother around the house by cooking and cleaning. As a young chef, it is essential to know where your ingredients come from. One of my favorite memories on the farm is making pizza because it was the first meal I got to make as a junior chef. I also love feeding the chickens because I like to feed people and animals. The farm has a positive impact on the community because it gives people something to do and keeps them busy. Without the farm, people in the neighborhood could potentially do harmful things. The farm is an example of the positive that can be done in the community and the kids see that. It shows the youth they can create their own path, go where they want, and not stay stuck in their surrounding environment. Youth like myself love the farm, we like to plant vegetables and watch them grow as we grow too.

Samia Rab Kirchner: Circle of Influence and Cycles of Nature - Urban Farms and Urban Design

As told by Curtis Cherry

My life has been intriguing and dynamic. I have encountered many strangers who have transformed me. Most recently, it has been my experience at Plantation Park Heights Urban Farm. I have gained experience growing plants with Farmers Chippy and Tiara, the leaders of the farm. Learning about growing plants brought me to a full circle moment when I remembered living in my small town Quetta in the South Asian country of Pakistan. Several family members believed I was going into agriculture when I was actually going to school for architecture. Years later, I ended up co-designing a demonstration kitchen for the farm, blending the two worlds. It was a joy to co-design a demonstration kitchen during the pandemic when food access was very limited for Park Heights residents.

Before encountering Plantation Park Heights, I was a wandering woman educating future architects on four different continents. I moved to the United States about thirty-five years ago. Teaching at Morgan State University is how I initially connected with Farmer Chippy in the spring of 2020. Since then, I have centered my urban design studio classes around the farm. Doing so was a breath of fresh air because it allowed me to continue teaching outside during the pandemic rather than hosting Zoom classes remotely. Since then, six environmental design courses have focused on residents' access to healthy food and healthy neighborhood design. Even though I was teaching college level courses, I began learning from the youngest farmers who helped me learn from nature. You are never too old to let nature teach you. Once we were outdoors we got to learn from the trees and I used them as a metaphor for design. I showed students how trees self-replicate their simple design of the leaf to create a complex ecosystem.

Like me, Farmer Chippy left his home in Trinidad, where there is a significant number of people of the South Asian Diaspora. He burns Bakhoor, a form of incense, at the farm. The smell of Bakhoor is a common thing we both grew up with. The smell brings memories of home and perhaps this is why the farm feels like home. Growing up I loved eating dahl (lentils) as it was a staple food for all my childhood

66

meals. Farmer Chippy prepares it often at community cooking events and I just love it! Everyone should give it a try. It is delicious and a good protein substitute for animal meat. Don't get me wrong, I love my goat and lamb, I could never envision giving them up. Yet at the farm, I am exposed to the freshness of vegetables that offer an alternative to my daily diet. This is a glimpse of how immigrants can find culturally relevant foods, with maximum health and environmental benefits, at urban farms. The farm often attracts people from the Caribbean, African, Latin American, and South Asian diasporas. As human beings, we should be flexible in our ways and allow strangers to shape our behaviors and daily routines. We have to enjoy our own personal growth and surround ourselves with people who push us to grow on a daily basis. It's about creating a sense of belonging within our communities.

Growing up, I knew I liked to use my imagination and thinking skills to make a difference in the world. I used to be interested in textile design. However, I preferred analytical drawing over free hand drawing which is why I moved into architecture. I knew I had a purpose but at the time, I didn't know what that purpose was. The search for creative purpose has helped me find my natural rhythm at Plantation Park Heights. Planting and watering seeds is important, but there is more to the practice of farming than that. There are plants that the farm can only grow from April to November. At any other time of the year, there needs to be a means of keeping food available for people to eat if they are going to rely on farming as a main source of sustenance. This annual challenge for farmers is why processing and preserving produce is an important skill we are prioritizing.

I have great appreciation for Plantation Park Heights because here I learned from the natural cycles of life. There is a simplicity to how plants grow and germinate while providing food for people. Since we eat plants and they have beneficial nutrients, we need to be part of their process. As I connect with plants, I recognize that it is important to grow and harvest *ideas* from diverse encounters, making our designs relevant to the next generation. At the farm they use fresh seeds to grow the next cycle of plants. The plants that grow on the farm tend to increase in size and are more enriched after the first time they get planted. In urban design, germinating an idea based on existing ideas that have flourished in the neighborhood can be a tool that improves critical thinking. I feel like I'm in my element when there is a difficult situation that needs a creative design solution. The farmers at Plantation Park Heights present me with challenges in ways that shape my theory of urban design.

Design has played a critical part in shaping Baltimore's segregated neighborhoods, where access to healthy food fluctuates with the color of residents' skin. Northwest Baltimore residents of Park Heights have limited options in accessing healthy food, leading to the district's status as a "food priority area" in Baltimore City. Many residents have taken to urban farming as an act of living with dignity. While Fox News celebrates Baltimore's high crime rate and low performing schools, Plantation Park Heights Urban Farmers are feeding residents, continuing the tradition of black liberation movements, and fighting for equity through the armor of healthy food access.

Plantation Park Heights Urban Farm is expanding the potential of Park Heights being food secure. Their circle of influence is increasing, as reflected in the zip codes from where people are coming to pick up free food boxes on Thursdays. The farm continues to employ a cohesive urban design that effectively utilizes all the open spaces in its custodianship. We, as experts, have an opportunity to engage the experiences of urban farmers to theorize an emerging focus for urban design. We seek increased access to green spaces while making communities' food secure and designing for youth to experience nature. Plantation Park Heights is part of my journey of how I got to where I am today. If I were to collect the zip codes of where all my students are impacting communities through their designs, the map would probably cover most of the world. This is a circle of influence that I hope perpetually grows as I learn from my students who are now engaging the farm as their site of transforming Baltimore neighborhoods.

PART III: RESPECT

SHAWNA CHEATHAM: UNIQUE LIKE SNOWFLAKES: NEURODIVERSITY IN URBAN FARMING

AS TOLD BY CHIDERA NDUBUEZE

I cannot emphasize the importance of spaces like Plantation Park Heights Urban Farm, especially for people like me.

I am a Baltimore native, born in a home on Shirley Avenue in Park Heights. I came to know Plantation Park Heights after my first encounter with Farmers Chippy and Tiara at Be More Green. They were there visiting Farmer Nell of City Weeds, who I've been partnering with for three years. I instantly formed a sisterhood bond with Farmer Tiara. They invited me to Plantation Park Heights and I went to visit them there about a week later.

My first visit to the farm changed my life. It was a cold February day, shortly after my solar return. All of us who were working that day seeded plants until we were exhausted, then we sat around the fire and got into a rap cypher, which is a freestyle circle. We ate some of Farmer Chippy's famous soup and talked through the night as the two shared the history of the farm, and how it started from nothing. Farmer Chippy told us about the Agrihood youth program and the wisdom they share with young people in Baltimore. We didn't even notice how cold it was because the energy was so warm. I loved the purposeful and fruitful conversations we had, and I instantly felt a sense of community and safety. The people at the farm just loved on me and I haven't been the same since. Farmers Tiara and Chippy became my sister and brother. Everyone gets to know each other and is part of the farm family.

Eventually, I brought my children and tribe to the farm. My children, friends, and I have learned a lot about urban agriculture from the Plantation Park Heights farmers, but the knowledge we've gained goes beyond farming. I have five children and all of them are neurodivergent like me. Three of my children are adults, and several of us are on the autism spectrum. We also have other, related diagnoses, which I call our alphabet soup. The mental health system and textbooks didn't help us much, but at the farm we've learned about ourselves, and have connected with the earth, our culture, and our original nature.

I initially brought one of my children to the farm because I knew he needed grounding. Even though he had been reluctant to come, he instantly bonded with Farmer Tiara, and now he has a strong connection to the earth. Whenever I can get my youngest to the farm, he doesn't want to leave! Still, environmental factors like mainstream media make it inevitable for unwanted seeds to be planted in the psyches of our children. They need a safe, peaceful, flourishing, and genuine space to combat what people assume is typical here in Baltimore. Adults are welcome, but the farm's mission is to be a safe space for the youth, for the corner boys and others who are left out. There is no violence here and people are free to be themselves. My first born is a victim of the school to prison pipeline, he'll be home soon and this situation reminds me why these spaces are important. Like the plants, the trees, and all things in the soil, each human being is distinct. All individuals on the spectrum are unique, like snowflakes. That is what makes all of us sprout beautifully. We complement each other. That is what a garden is.

I have always been fascinated with nature and my environment, and now, more than ever, urban farming. You'll often find me barefoot, grounded in the earth, with my hands in the dirt, experiencing the different textures, temperatures, and colors. I encourage everyone to take their shoes off, to release themselves from artificial compounds, and get back into their original, indigenous state of being. I don't think it takes even five minutes before our systems begin to recalibrate, because the soil has components that facilitate our healing.

Urban farming is calming and therapeutic. The free-flowing, organic, safe space at the farm has become a huge part of my continued healing journey. For people like me, with autism and epilepsy, it is vital to go back to nature and sustain ourselves by growing our own food. It gives me an excuse to come outside. I don't usually like to go outside because I can feel everything, and it often makes me feel imbalanced. I typically stay at home unless there's an intentionally safe space for me to go to, like the one at the farm. I am at my most social at Plantation

Park Heights and Be More Green. No matter how bad the darkness in this city is, I have found a safe place with them where I can contribute to my community.

At the farm, I have learned how to identify various herbs, plants, and parasitic insects, who I refer to as "the opps." The farmers taught me how to take care of plants and how to seed trays of basil, kale, and sunflowers. In fact, most of the plants at Be More Green are the grandbabies of Plantation Park Heights plants. I have grown the most beautiful sunflowers, courtesy of Farmers Chippy and Tiara, in our urban garden at Be More Green. Making these connections is part of the work of my nonprofit, AFRO Aspies ROCK. We connect neurodivergent and neurotypical communities in creativity, community engagement, service, and healing arts workshops while addressing stigmas about people on the autism spectrum. My children and I facilitated our first community garden planting day for AFRO Aspies ROCK using the knowledge and skills we gained at Plantation Park Heights.

Farm work is not for the weak. We get our hands dirty. Even though my nails and fingers are adorned, I don't care much about that kind of decoration. What really matters is underneath our feet. This is the cycle of life: we give to the earth and she gives right back to us. What she gives me through this free-flowing, organic, safe space is a healing journey with everything in my life integrally and intricately connected. This work is so important for our wellness and healing. Some may ask "Aren't you already healed?" They fail to understand that healing is not a linear process, it is a continual journey. Urban farmers like Tiara, Chippy, and Nell are helping educate people about how food facilitates their wellness. I even make holistic teas through my brand SheBREWS that help with digestive and reproductive health. I live and breathe this stuff! This is a cycle of love, generosity, and reciprocity. Through the patience of growing plants, I am learning to be patient with myself and other people.

I like to think of farm work as a process of birth and death. It's always sad when plants die, but I know that they go back to the earth so that something else can be born. Energy does not die, it only transitions. Urban farming has helped me understand the alchemist that I am and that we all have the potential to be. The skills I've learned at Plantation Park Heights with my children, who I consider to be my divine starseed, have saved my life. I give this space the highest praise and honor. It is a sacred place. As a neurodivergent person, I often struggle with the challenges that come with being on the spectrum. Sometimes I get frustrated with what I cannot do but when I see my plants grow and flourish it brings me a sense of peace. I say to myself, just as these plants are fighting and flourishing and surviving, so shall I, so will I, so am I. I am Plantations Park Heights Urban Farm, I am Be More Green City Weeds, I am AFRO Aspies ROCK, and I am Shawna Divine.

KHALIA YOUNG: AN ABUNDANCE OF FOOD

AS TOLD BY MAYA TOLAND

I'm from Cincinnati and I went to the University of Toledo
for undergrad, majoring in Public Relations (PR) and cultural
anthropology. I had a fellowship with NextGen Climate Action and
worked in the community, canvassing, registering voters, and hosting
conversations about the impacts of climate change. The people in the
community were super interested because they never had anyone,
especially a young Black woman, engage with them. This shifted my
mindset from doing PR and corporate work to community work.

I reflected on my interests and began researching different careers
online. Eventually, I came across City and Regional Planning, which
I had never heard of. Within City and Regional Planning there's
transportation, sustainability, community development, food, design,
and more. I started to research the highest paying city planner jobs
in the United States. Washington DC was on my radar but it's super
expensive to live there so I looked to see what was close by. I found
Morgan State University and ended up in the City and Regional
Planning Master's program at Morgan in Baltimore City. The program
consists of rigorous planning and critical thinking to develop land use
and create successful communities for people to live in. I had only
been to Baltimore once but when I told people I was thinking of
moving here, they were concerned. However, I was indifferent and
open to creating my own experiences. I didn't come here afraid and
when I got here, I loved it.

The goal of city and regional planning is to develop areas that
have efficient economic and social development, catering to the
needs of a community. Throughout my experience at Morgan, I was
involved in many community projects and worked with professors
like Dr. Kirchner. Through her studio classes, I was able to connect
with Farmer Chippy at Plantation Park Heights Urban Farm. For my
project, I held focus groups at the farm to see if the Department of
Planning's community engagement sessions included residents across
communities. Many residents didn't even know about the sessions,
proving the city probably didn't do much on the ground when trying to
find community members. They most likely went through faith-based

groups, nonprofits, and community associations that already had engagement. I used the Park Heights residents' feedback to offer the city recommendations on how to be more inclusive and engaging.

Prior to living in Baltimore, I had never experienced a lack of access to healthy food because grocery stores were within walking distance. Park Heights is considered a food desert. But the term food desert is interchangeable with *food apartheid*. Food desert implies that it's naturally this way. Neighborhoods are not just naturally lacking food. Food apartheid speaks to the intentional structure of a community. When you change the language, you hold people accountable for why things are the way they are. Who is doing this? Why is it this way? When you use words like "apartheid" and "sovereignty" you give the space for accountability and solutions. This is what the farm is up against, the systemic racism and planning policies that have created a neighborhood lacking in resources.

Farmers Chippy and Tiara are doing a lot of work in food sovereignty as they take their power back and control of their own food. They are not depending on grocery stores or someone else to feed them. They are connecting with the land and feeding themselves. Plantation Park Heights is a thriving marketplace in the middle of a food apartheid that is providing the community access to good quality food and food security. Not only does the farm give out food, but they also provide cooking lessons, recipes, and educational opportunities. Lack of food access can also mean lack of knowledge on how to properly prepare different kinds of fruits and vegetables. The mission of the farm is to introduce people to healthier lifestyles, to raise young farmers, to incorporate leadership skills into agriculture, and to build relationships. I believe that Plantation Park Heights has absolutely developed a solution to solving food apartheid in Park Heights. The impact of the farm is changing the narrative of the community for the better.

Experiencing this type of atmosphere in such a short amount of time has encouraged me to keep visiting. I have been at the farm for three years now but I have been consistently volunteering for about a year. I've chosen to prioritize the farm. I've gone to my last two employers and let them know that I plan to continue volunteering at the farm. I've introduced my former employer, a real estate developer, to Farmer Chippy and they're working to incorporate urban farming into his residential developments. Urban agriculture is new to me because I didn't grow up with conversations about what food does to your body or how to grow your own. It was refreshing to have these conversations at the farm and see people engaging children in this work. The farm is very inclusive and intergenerational: there's elders, adults, teenagers, and children. Within these different age groups, you have people from across the city. You have the corner boys, corporate people, school leaders, and artists. I love that this is not your typical farm.

Plantation Park Heights has impacted me immensely. My all-time favorite thing to do at the farm is meeting and conversing with diverse groups of people who expand my knowledge of Baltimore City. Before making this life-changing decision to move to Baltimore, I always felt like I was connected to the earth. I spent a lot of my childhood at the park, playing in the woods and in the dirt, but I never engaged through food. Now I'm connected to the earth because I've learned how to harvest. I've planted tomato seeds and experienced the physical aspect of planting those seeds. It has opened my eyes to urban agriculture and changed my perspective on farms. The farms in Cincinnati are mostly acres of plain corn fields or flatlands and I thought that was the only definition of what a farm was. Coming to Baltimore and volunteering at Plantation Park Heights, I now see there is much more to what a farm can be.

I am elated to be a part of the farm's journey and to watch it evolve. One asset I can bring to the farm, to assist in its success, is bringing more people and putting the word out there. I am heavily interconnected with several organizations and do outreach through my own community engagement consulting company, The Key to Planning. My mission is to keep people connected and aware about what is happening in their communities, eventually including them in the development process to ensure their voices are heard. I want to build a partnership between the farm and The Key to Planning to support in their advocacy work, training, and programming. I'm excited to continue inviting my sorority sisters and friends to the farm, changing their outlook on urban agriculture. For Founders' Day, my sorority, Alpha Kappa Alpha, volunteered and built raised beds. I even had a friend start to volunteer at the farm who's helped plant different vegetables during the spring planting season. My vision of my relationship with the farm includes me having more of an impact. I'm not sure what that looks like right now but I always keep the farm in mind when it comes to anything I'm doing because of its major role in my life.

Harold Morales: Choosing How Much of Yourself to Bring With You

As Told By Chandler Bell

I am a lifelong student, learning from mistakes and changes, from love and fear. Despite all of it, I'm still here, figuring out how to move more gracefully through the world. My family's roots are in Central America and I was born and raised in Los Angeles. As an adult, I've lived in Pennsylvania and now I'm in Baltimore. The ground beneath me never seems to completely settle. In this state of disorientation I've grown accustomed to, I've realized that when on a journey, you have to choose how much of yourself to bring with you. You move into a space and can bring parts of your previous self with you, translating it to the new context, or you choose to let it go. I like wearing different hats when traveling, literally and figuratively. I have fun getting lost searching in a closet full of possibilities for what feels right on any given occasion.

In 2020, I visited Plantation Park Heights Urban Farm as director of the Center for Religion and Cities (CRC) at Morgan State University. The CRC had been organizing a conference for April of 2020 to connect academics, activists, and artists in Baltimore City. We canceled the conference in the wake of the COVID-19 pandemic. We asked permission from our sponsor, the Henry Luce Foundation, to shift the funds from the conference to direct aid. Not only were they supportive, they also provided additional funds to support our community partners' relief and restoration efforts. In addition to providing mini-grants for direct aid, we put together a team of thirty researchers to document and reflect on the work of our community partners. It was at this time that Dr. Samia Kirchner, a good friend and collaborator, connected the CRC with the farm to co-design an outdoor demonstration kitchen for Park Heights.

My first day at the farm was for the unveiling of the student designed demo-kitchen. Facing the farm's hoop house, there were picnic tables filled with people listening to the presentations. It was a hot and humid day. I had my favorite large hat on for shade. I remember the cicadas were especially loud then, which added to the apocalyptic feel of the times. It was then that I met Farmer Chippy and Farmer Tiara. Chippy

84

was also wearing a big farm hat and colorful socks. We hit it off right away!

They were introducing medicinal herbs to the community and teaching us about the importance of what we put into our bodies. I asked about the spiritual inspiration for the farm. They answered enthusiastically, speaking at length about the Ifa tradition that animated their work and dedication to the land and community. The spirit that animates the work is deeply connected to the ancestors of the people connecting with the earth here. Like me, people are connecting to a part of themselves they are too often estranged from. Continuously questioning your deepest beliefs about who you are, where you come from, and where you want to go is hard and uncomfortable work. Rather than shy away from this difficult but important work, Chippy and Tiara cultivate practices for tending to, shaping, and harvesting meaningful, spirit-filled agrarian work.

Lately, I've been reflecting a lot on one of Octavia Butler's *Earthseed* quotes from her novel *Parable of the Sower*: "All that you touch, You Change. All that you Change, Changes you. The only lasting truth is Change." Working on the farm, I've grown a deeper appreciation for the changes that the seasons bring and the anticipation and preparations that are possible. My ongoing prayer is for grace and to move more gracefully through change with my loved ones. Along with Chippy, Tiara, and others, I continue to find forgotten parts of myself while also dawning new hats, socks, and satchels. I create new tools and rituals for the survival and wellbeing of myself and communities in a world that, like me, is constantly flowing and moving.

Transient Being

Do you believe in the things you cannot see?

Some days I feel like I'm playing make believe,

My closet feels like a dressing room that I run in and out of
to change who I am for the things people around me need,

I am a master juggler in a traveling circus
that has uprooted my safety net.

And I wait to fall.

I brace myself for the impact and breathe.

And when I do,

It is inevitable that the people I love,

My chosen family,

Will be there to help me down softly.

I have experienced different

Things.

Places.

People.

And all of those things have experienced a different version of me

I can acknowledge that there are pieces of me that cycle in and out,
that there is a time and place for my different personalities.

I have found a part of me here though,

In the silence, in the middle of the city.

In the people that welcome me,

In the committed relationships I have created
with the seeds I've sown.

My energy is put into the circles of serenity
where I have left a part of me.

TJ Triplett: A Farm Like Family

As told by Moniesha Lawings

"Come here, let me show you around!" A woman called to us, with a wide smile.

My friend, my brother, and I play outside in our neighborhood almost every day, but one day, we wanted to see what was in an alley we had never gone down before. We weren't sure exactly what we were witnessing. Rows of plants, a shed looking thing, and the sounds of chickens? Then, a woman called us over. She introduced herself as Ms. T, and said she was a farmer. She showed us around the land and it was a farm right in the middle of our neighborhood. I was surprised. I had never thought about there being a farm in the middle of the city, but it was so cool! Ms. T told us to come back and help out whenever we wanted and we did. So yeah, about two years ago, I became a farmer too.

My name is Tyler, but most people call me TJ. I grew up here in Baltimore City with my family. So far, I feel like I have had a pretty fun childhood, surrounded by people who love and care about me. I love my neighbors and community; they always look out for me. The volunteers, farmers, community members, and everyone at Plantation Park Heights Urban Farm has become my family.

During the day, I attend Pimlico Elementary/Middle School. I think I'm a pretty good student and volunteering at the farm has definitely helped. At school, in my cooking class, I'm able to recognize the plants and vegetables we use in recipes because I learned about them when helping grow them at the farm. After school, I know I need to finish my chores and homework before I can go help out at the farm, so I feel like I'm more responsible and eager to do so. After getting everything done, I go to the farm and do whatever the farmers need help with: watering plants, carrying stuff, picking vegetables, anything. Overall, it's fun and I love getting to work with my friends and the other volunteers. I also appreciate having mentors around, like Mr. Chippy and Ms. T, who believe in me and help me stay out of trouble.

The farm is not only important to me, it's also important to the community because it helps out Park Heights. It gives people, especially kids, something to do, so they can stay out of trouble. We work hard, but we also get to have fun. My favorite thing we do is called Camp Night. We, the kids and other volunteers, come to the farm, set up tents, and get to spend the night outside. Every week we give out free boxes of food, helping people who need it. There are classes and other fun things to do, so people can learn how to grow things and get involved with the farm too. Lastly, this work is important for Black people because we don't see this a lot...Black people farming or being in nature. It's cool to see.

Before finding this farm, I didn't know anything about farming or urban agriculture, and I definitely didn't think about how big it could be for so many people. After volunteering here, though, I definitely do. There aren't a lot of things like this here in the city that are making a difference, so I really want to see other kids get into it too. Farming would give kids something fun to do, so they don't feel like they have to do bad things or be in the streets. At least try it out! You might find out that you like it.

I really love Baltimore. I know the things people say about it, but I want to be here forever. I don't really know what I want to be when I grow up, but I do know that I want to be here in the city to guide the next generation like Mr. Chippy and Ms. T. I want to give kids advice, help them find good things to do, and believe in them. I want to take over right here, in a positive way of course, so everyone can know that their futures have so many good possibilities. That's what I want my future and the future of Baltimore to look like.

FARMER CHIPPY: FROM TRINNY TO PARK HEIGHTS

AS TOLD BY TIFFANY ODURO

At first, the farm was merely a social test. Now, it has evolved into a legitimate company. I was born in Trinidad. After relocating to the neighborhood of Park Heights in Baltimore City, I realized that Maryland needed a taste of the Caribbean. I set out to make a difference and improve the lives of the adolescents and young adults who live in the Park Heights community. I realized that locals should be encouraged to cultivate their own food rather than purchase packaged and often unhealthier versions of the same food. The looks on young people's faces when they consume food that they assisted in growing is fantastic. I got to where I am with the farm through hard work and dedication. I ignored all the detractors, stayed focused, pursued my goals, and found some folks who were aligned with those goals. We empower each other and now here we are at this amazing farm for us, by us. Plantation Park Heights Urban Farm's mission is to offer wholesome food and community growth possibilities in Park Heights and beyond.

Trinidad is a very lovely island. It was dominated by oil and gas but, along with most of the residents, my family farmed and grew something. Those experiences shaped me because regardless of what I did in life, I always wanted to grow my own food. I continued with that passion and here we are. Growing up in Trinidad, we had a colonial influence. My parents were born when we were still colonized by the British, before Trinidad and Tobago became independent in 1962. I was born in the 70s so I'm the first generation free. Free thoughts, free motion, and free behavior was a revolution in the 70s. Trinidad as a country changed its negative perceptions of Black skin color and instead there was love and care for us Black people. I still feel very revolutionary. I feel like my leader at the time, Dr. Eric Williams, who always prided himself in saying that every creed is equal and that there is no place in a civilized society for bacchanal and confusion.

A moment in my life that altered and shaped my path was when I became a biomedical field service engineer. I was interested in healthcare, science, and cells so I looked for companies that grew cells outside the human body and separated proteins. Understanding these concepts taught me that healthcare and regenerative medicine can have preventive and personalized methods. If you get in front of it and start prioritizing preventive healthcare like growing food in your backyard and knowing what you eat, chances are, you can just meet with your physician and dentist for assessments twice a year, as opposed to waiting to get sick.

The thing that would cause me to leave Baltimore and relocate is the expansion and growth of my business. I would take these philosophies and technologies I've grown to third world countries to help less fortunate people across the globe. That is my mission. Before I get there I have to make jokes in every corner of the United States, especially places that are home to our African and Caribbean diasporas, like Detroit and Chicago. We've got opportunities to expand what we've discovered or rediscovered here in Baltimore and I want to do that globally. I'm rolling now as the farm's Executive Director but this role will probably only last for the rest of 2023 because my intention is to take this to every corner of the United States. I hope to one day make this international, for third world countries in particular. Governments need to do more for those who were left out *prior* to COVID.

I have so many great memories of these young, talented farmers from Park Heights. One of my favorite memories is when Tevin first came to the farm, not knowing anything about peppers. We put him to work and he planted peppers with us, covering the entire hill. By the time the work was done he was able to name twelve peppers. He wasn't even ten-years-old back then. A few years later, he's still here now, teaching even younger generations to farm. He's just one of the many children who blow me away. I have a 120% commitment to Plantation Park Heights Urban Farm. I cannot abandon this community because we started this farm for them. I like to say, "We got it out the mud," so I'll always be here. Most of the people I'm training will continue to innovate and improve the quality of life for our communities. This is a springboard for talented people. I want to mentor them and direct them to places like the United States

Department of Agriculture (USDA) and the Maryland Department of Agriculture because I want to grow the next generation of farmers.

Why am I at the farm? I am here to distribute healthy food while empowering young Black and brown children to grow their own food. Every Thursday, we give away fresh fruits and vegetables that arrive on pallets delivered by The Common Market's Farm-Fresh Box Program. We do cooking demonstrations where we prepare, cook, and serve soups made up of whatever is in the boxes that we give away. We do this to show people how to prepare it. They learn, they eat, and they say, "Yum!" I'm going to replicate that too.

Young people in the region have an alternative option to drug use and criminal activity by participating in urban farming. The farm's crops are planted, grown, and harvested by volunteers who live nearby. It offers them methods to not just survive, but to thrive. Contributions to the farm have not been ignored. The fact that there are citizens like us who take pride in their community and care about their neighbors makes us farmers in Park Heights very lucky. I'm appreciative of everything we've done and I'm glad to be at the farm.

My name is Richard Francis, but everyone calls my Farmer Chippy, and I work at Plantation Park Heights Urban Farm.

21 Questions with Farmer Chippy

1. **How has the farm contributed to your spiritual life?** It's all one big blur. It's all one, together. There's no separation between myself and my spirituality and this farm. It's all connected.

2. **What are your core values and where do you see them at**

the farm? Love, loyalty, and respect. It is our mantra. This is a place where everyone can come. It is a place for everyone. Free of fear and free from fear.

3. **How did you meet Farmer Tiara? What has it been like growing together?** I met her on the block in Baltimore City, she's a real Baltimore girl. I told her my idea and she said, "Aight, let's go" and we're here now. She never turned back.

4. **If you could give young people in the world one piece of advice, what would it be?** Find yourself and believe in that. Find yourself and believe in that thing that you found.

5. **If you could give old people in the world one piece of advice, what would it be?** Thanks for coming, now listen. Let us show you how to do it.

6. **What do you think your family, ancestors, and mentors would think of you now?** From their mouths, in the last three weeks, I saw two mentors and they said to me, "I don't know what you did in the past, I have no clue what's going to happen in the future, but I know you and you're living your best life."

7. **How do you encourage creative thinking at the farm?** Allow everyone to express who they are and what they do without fear.

8. **What would you say your superpower is?** Bringing the best people on the planet together to figure out solutions for the future.

9. **What is the most important characteristic of a leader?** Patience.

10. **What mistake do you see leaders make frequently?** Expecting the others to perform as good as they are.

11. **What's the most important risk you took and why?** I believed in myself because the others disappointed me.

12. **What are your hopes for the future of urban agriculture?**
That Baltimore will be a place with beautiful trees, clean food,
great artists, great musicians, wonderful teachers, trainers,
and community members. A place we can call our own. A
place where there's a thriving market, community shared and
supported agriculture, and an urban farmer training resource
center right here with all the extension offices based right here
in Park Heights.

13. **What misconceptions about farming would you like to
dispel?** Farming is hard work. Farming is not hard work, it is
fun work.

14. **What motivates you to keep going even when times are
tough?** Watching my children develop into wonderful farmers
and knowing that I cannot stop now that I've brought them
into this fray.

15. **What are your current priorities in this season of your
life, and why?** Continue to grow cleaner, greener foods. Feed
the nation as part of my commitment to ensure each child
understands where their food is coming from. So I want to
continue doing that work to make sure children know where
their foods are coming from.

16. **What makes your life feel purposeful?** Waking up in the
morning.

17. **Describe your perfect day.** A day where everyone on the
team is smiling and we are together - that's the perfect day.
Everyone on my team is smiling and we're together.

18. **What's one of your favorite childhood memories?** Going
to the market with my mom, having a good time with the
people in the neighborhood, and everyone calling me Darling.
All the older ladies saying, "Hello darling." I remember that.

19. **How do you know when you can trust someone?** Oh. Oh,

that's difficult. I mean, I don't take care of that, the God I serve takes care of that. I don't trust people really, I trust spirits. I don't really mess with humans, humans are...human beings are not trustable so I trust the spirit that dwelleth in the moment that it dwells. So if you come around me and you're in happy spirits, we good. If you come around me and you're not in happy spirits, I wanna know what's wrong. We ain't good cause if you ain't good, I ain't good.

20. **What's your favorite food?** Oh, that's easy, bake and shark.

21. **If you were stranded on a deserted island, what three things would you bring with you?** A lighter, a machete, and...hmm...that last one's challenging. A blow up raft.

DISCUSSING THIS TOPIC IN YOUR COMMUNITY

Because The Facing Project is steeped in empathy and connecting across differences, Listening Circles are a fantastic way to bring more people into the conversation on the topics/themes addressed in this book.

Listening Circles provide readers with a more intimate experience to reflect upon the stories and ask questions in a controlled environment. However, it's important to include a trained facilitator who understands how to moderate, when to let conversations flow, and when to step in to move them along.

If you choose to include Listening Circles in your community, we recommend having at least four different locations and dates for these to happen, and be sure to have folks register in advance. Hosts could include area libraries, schools, and/or colleges and universities.

Also, it's important to set the following standards at the beginning of each discussion:
    1. We acknowledge that we are all here to learn with open hearts and open minds.

2. Before speaking or asking a question, we all agree that we will take a moment to reflect on if "my voice/question matters in the particular moment" or if "I should give the opportunity to someone else to speak."

3. R-E-S-P-E-C-T is more than an Aretha Franklin song; respect is a value that we will hold close throughout our discussions.

4. We honor the storytellers who shared their experiences, and our goal is to not discount them but rather to understand how all of our stories are intertwined and part of the human condition.

Ideally, Listening Circles should have no more than 20 participants in each circle. If you find that one of your locations may have more than 20, you'll want to explore breaking them up into more than one group.

Also, participants should have read a copy of this book before participating. This makes for deeper discussion, and it keeps the facilitators from having to give a full breakdown of all of the themes/topics included throughout the book. More copies can be ordered at www.facingproject.com.

However, it's always good to open a Listening Circle with a reading of one or two stories that immediately follow introductions and community standards. Ask for one to two attendees to volunteer to read the selected stories aloud to the Circle.

Then have the facilitator ask the group: How did those two stories make you feel?

And be sure to have them follow-up with "tell me more" and other open-ended questions. Of course, a trained facilitator will understand how to let this process flow.

Lastly, be sure to include action items for the participants. This could include other events in your community centered around the topic/issues addressed in this book, volunteer opportunities with nonprofits, and/or other ways they can get involved.

Sponsors

ABOUT THE FACING PROJECT

The Facing Project is a 501(c)(3) nonprofit that creates a more understanding and empathetic world through stories that inspire action. The organization provides tools and a platform for everyday individuals to share their stories, connect across differences, and begin conversations using their own narratives as a guide. The Facing Project has engaged more than 7,500 volunteer storytellers, writers, and actors who have told more than 1,500 stories that have been used in grassroots movements, in schools, and in government to inform and inspire action. In addition, stories from The Facing Project are published in books through The Facing Project Press and are regularly performed on The Facing Project Radio Show on NPR.

- Learn more at facingproject.com.

- Follow us on Twitter and Instagram @FacingProject, and on Facebook at *TheFacingProject*.

Milton Keynes UK
Ingram Content Group UK Ltd.
UKHW040708201123
432908UK00001B/188